HOW TO MASTER HOME PLUMBING

The Ultimate DIY Guide to Fixing Leaks, Clogs, and Common Plumbing Issues with Step-by-Step Instructions, Expert Tips, and Proven Techniques for Homeowners

The Fix It Guy

Chapter 1
Understanding Your Home's Plumbing System

The Anatomy of Your Plumbing: Pipes, Fixtures, and Valves

Before you can start tackling plumbing projects and repairs, it's essential to have a solid understanding of how your home's plumbing system works. In this chapter, we'll dive into the anatomy of your plumbing, exploring the various components that make up this complex network of pipes, fixtures, and valves. By the end of this chapter, you'll have a clear picture of how water enters your home, circulates through your fixtures, and exits as waste.

Your home's plumbing system consists of three main components: pipes, fixtures, and valves. Each component plays a crucial role in ensuring that water flows smoothly and efficiently throughout your home.

Pipes:
The pipes in your plumbing system are responsible for transporting water from the main water supply to your fixtures, and then carrying waste water away from your home. There are two main types of pipes:

1. Supply Pipes: These pipes bring clean water into your home from the main water supply. They are typically made of copper, PEX (cross-linked polyethylene), or CPVC (chlorinated polyvinyl chloride). Supply pipes are under constant pressure to ensure that water flows readily when you turn on a faucet or flush a toilet.

2. Drain Pipes: These pipes carry waste water and sewage away from your home and into the main sewer line or septic tank. Drain pipes are usually made of PVC (polyvinyl chloride), ABS (acrylonitrile butadiene styrene), or cast iron. They are larger in diameter than supply pipes and rely on gravity to move waste water out of your home.

Fixtures:

Plumbing fixtures are the devices in your home that dispense and control water flow. Some common fixtures include:

1. Faucets: Found in kitchens and bathrooms, faucets control the flow of water for tasks like washing hands, cleaning dishes, and brushing teeth.

2. Toilets: Toilets are essential fixtures that remove human waste from your home. They consist of a bowl, a tank, and various internal components that work together to flush waste into the drain pipes.

3. Showers and Bathtubs: These fixtures provide a place for bathing and rely on a combination of supply pipes for hot and cold water, as well as drain pipes to remove waste water.

4. Sinks: Kitchen and bathroom sinks are used for a variety of tasks and are connected to both supply and drain pipes.

5. Appliances: Dishwashers, washing machines, and water heaters are also considered plumbing fixtures, as they are connected to your home's water supply and drain system.

Valves:

Valves are devices that control the flow of water through your plumbing system. They allow you to shut off water to specific fixtures or sections of your plumbing for repairs or maintenance. The most common types of valves in a home plumbing system include:

1. Main Shut-Off Valve: This valve controls the main water supply to your entire home. It is usually located near the water meter or where the main water line enters your house.

2. Supply Valves: These valves control the water supply to individual fixtures like toilets, sinks, and washing machines. They allow you to shut off water to a specific fixture without affecting the rest of your plumbing system.

3. Pressure Reducing Valves: These valves reduce the water pressure entering your home to a safe level, protecting your plumbing fixtures and appliances from damage caused by high water pressure.

4. Check Valves: These valves allow water to flow in one direction only, preventing backflow and potential contamination of your water supply.

Understanding how these components work together is the first step in becoming a confident DIY plumber. In the following sections, we'll explore each component in more detail, giving you the knowledge you need to diagnose and repair common plumbing issues in your home.

Water Supply Systems: From the Main Line to Your Faucet

Now that you have a general understanding of the main components in your home's plumbing system, let's take a closer look at how water makes its way from the main supply line to your faucets. In this section, we'll trace the path of water through your home's supply system and discuss the key components that ensure a steady flow of clean, safe water.

The Main Water Supply Line:
Your home's water supply journey begins at the main water supply line. This line, which is typically owned and maintained by your local water utility company, carries water from the municipal water source or well to your property. The main supply line is usually made of heavy-duty materials like copper, ductile iron, or PVC, and is designed to withstand high water pressure and underground conditions.

The Water Meter:
As the main supply line enters your property, it passes through a water meter. The water meter is a device that measures the amount of water your household consumes, which is then used by the utility company to calculate your water bill. The meter is typically located near the property line or in a utility box near your home.

The Main Shut-Off Valve:
Just after the water meter, you'll find the main shut-off valve. This valve controls the flow of water into your home and is crucial for emergency situations or major plumbing repairs. It's important to know the location of your main shut-off valve and to ensure that it is easily accessible and in good working condition.

Supply Lines:
From the main shut-off valve, water enters your home through a series of supply lines. These lines are typically made of copper, PEX, or CPVC and are smaller in diameter than the main supply line. The supply lines branch out to distribute water to various parts of your home, such as the kitchen, bathrooms, and laundry area.

Pressure Reducing Valve (PRV):
In some homes, particularly those in areas with high water pressure, a pressure reducing valve (PRV) is installed on the main supply line. The PRV reduces the incoming water pressure to a safe level, typically between 40 and 80 psi (pounds per square inch). This helps protect your plumbing fixtures and appliances from damage caused by excessive water pressure.

Water Heater:
Before reaching your faucets, a portion of the cold water supply is diverted to your water heater. The water heater, which can be tank-type or tankless, heats the water to a set temperature and stores it for use. Hot water from the heater is then distributed through a separate set of supply lines to fixtures like showers, bathtubs, and sinks.

Fixture Supply Lines:
The final leg of the water supply journey occurs through the fixture supply lines. These lines, which are typically flexible and made of braided stainless steel or plastic, connect the supply lines to individual fixtures like faucets, toilets, and appliances. Each fixture has its own shut-off valve, allowing you to control the water flow to that specific fixture without affecting the rest of the plumbing system.

Faucets and Fixtures:

Finally, water emerges from your faucets and fixtures, ready for use. Faucets and fixtures are designed to control the flow and temperature of water, as well as to provide a convenient and aesthetically pleasing point of use.

By understanding the path water takes from the main supply line to your faucets, you'll be better equipped to diagnose and repair issues that may arise in your home's water supply system. In the next section, we'll explore the drainage system and how it safely removes waste water from your home.

Drainage Systems: How Waste Water Leaves Your Home

Just as important as the water supply system is the drainage system, which is responsible for safely removing waste water from your home. In this section, we'll explore the components of the drainage system and how they work together to ensure that waste water is efficiently and hygienically transported away from your living spaces.

Fixture Drains:

The drainage system begins at the plumbing fixtures in your home, such as sinks, showers, bathtubs, and toilets. Each fixture has a drain that connects to a trap, which is a U-shaped section of pipe designed to hold water and prevent sewer gases from entering your home. The trap also catches small objects that may have accidentally fallen into the drain, preventing clogs deeper in the system.

Branch Drain Lines:

From the fixture drains, waste water flows into branch drain lines. These lines, which are typically made of PVC or ABS plastic, are larger in diameter than supply pipes and are designed to carry waste water from multiple fixtures. Branch drain lines connect to vertical pipes called stacks, which carry waste water down and out of your home.

Stacks and Vents:

Stacks, also known as soil stacks or waste stacks, are vertical pipes that run from the bottom of the drainage system up through the roof of your home. They are responsible for carrying waste water from the branch drain lines to the main sewer line or septic tank. Stacks also play a crucial role in venting the drainage system.

Vents are pipes that connect to the stacks and extend through the roof of your home. They allow fresh air to enter the drainage system, balancing the air pressure and preventing water from being siphoned out of fixture traps. Properly functioning vents also help prevent clogs and slow drains by allowing waste water to flow freely through the system.

Cleanouts:
Cleanouts are access points installed in the drainage system that allow for the removal of blockages or clogs. They are typically located near the base of stacks, at the end of long horizontal drain runs, and at bends in the piping. Cleanouts have removable caps that can be unscrewed to allow access to the inside of the pipe for cleaning or inspection.

Main Sewer Line:
The main sewer line is the final component of your home's drainage system. This large pipe, usually made of clay, concrete, or PVC, carries waste water from your home to the municipal sewer system or your septic tank. The main sewer line is typically buried underground and has a slight slope to allow waste water to flow by gravity.

Septic Systems:
In areas where municipal sewer systems are not available, homes rely on septic systems to treat and dispose of waste water. A septic system consists of a septic tank, where solid waste settles and decomposes, and a leach field, where liquid waste is filtered and dispersed into the soil. Regular maintenance and pumping of the septic tank are essential to keep the system functioning properly.

Understanding your home's drainage system is crucial for maintaining a safe and hygienic living environment. By knowing how waste water is removed from your home, you'll be better prepared to tackle clogs, leaks, and other drainage issues that may arise. In the following chapters, we'll dive into specific plumbing problems and provide step-by-step instructions for resolving them.

Chapter 2
Essential Tools and Materials for DIY Plumbing

Must-Have Plumbing Tools for Every Homeowner

Before you can start tackling plumbing projects and repairs, it's important to have the right tools and materials on hand. In this section, we'll discuss the essential tools and materials every homeowner should have in their plumbing toolkit, focusing on the must-have items that will help you complete most DIY plumbing tasks with ease.

1. Adjustable Wrenches:
Adjustable wrenches, also known as Crescent wrenches, are versatile tools that can be used to tighten or loosen nuts, bolts, and fittings of various sizes. A set of two wrenches, typically in 6-inch and 10-inch sizes, will cover most household plumbing needs. Look for wrenches with comfortable handles and a smooth adjusting mechanism for easy use.

2. Pipe Wrenches:
Pipe wrenches are designed specifically for gripping and turning pipes, making them essential for tasks like tightening or removing threaded pipe connections. A set of two pipe wrenches, in 14-inch and 18-inch sizes, will provide the leverage needed for most household plumbing jobs. Make sure to choose wrenches with durable, serrated jaws for a secure grip on pipes.

3. Basin Wrench:
A basin wrench is a specialized tool designed for working in tight spaces, particularly when installing or removing faucets. need to complete your plumbing projects successfully.

It features an adjustable jaw that can grip nuts at various angles, as well as a long, pivoting handle that allows you to reach difficult-to-access areas behind sinks and toilets.

4. Plungers:

Plungers are a must-have for clearing clogs in sinks, showers, and toilets. There are two main types of plungers: cup plungers and flange plungers. Cup plungers are best for sinks and showers, while flange plungers, which have a built-in flange that fits into the toilet bowl drain, are designed specifically for toilets.

5. Augers (Plumbing Snakes):

Augers, also known as plumbing snakes, are flexible, coiled cables that can be inserted into pipes to remove clogs that plungers can't handle. Hand augers are suitable for most small clogs in sinks and showers, while closet augers, which have a longer cable and a specialized head for navigating toilet traps, are designed for clearing toilet blockages.

6. Hacksaw:

A hacksaw is a versatile cutting tool that can be used to cut through pipes, bolts, and other metal components. Choose a hacksaw with a sturdy frame and a comfortable handle, and make sure to have a supply of replacement blades on hand.

7. Tubing Cutter:

A tubing cutter is a specialized tool for making clean, precise cuts on copper pipes. It features an adjustable cutting wheel that scores the pipe as you rotate the cutter around it, ensuring a straight, burr-free cut every time. This is essential for creating leak-free connections when working with copper piping.

8. Pliers:

A set of pliers is indispensable for gripping, twisting, and pulling various plumbing components. Channellock pliers, slip-joint pliers, and needle-nose pliers are all useful for different tasks, such as tightening connections, holding small parts, and reaching into tight spaces.

9. Plumber's Torch:

A plumber's torch, also called a propane torch, is used for soldering copper pipes and fittings. It provides a concentrated flame that quickly heats the metal, allowing you to create strong, leak-free joints. Make sure to choose a torch with an adjustable flame and a sturdy base for safe and easy use.

10. Safety Gear:

When working on plumbing projects, it's crucial to protect yourself from potential hazards. Essential safety gear includes safety glasses to shield your eyes from debris and splashes, work gloves to protect your hands from sharp edges and hot surfaces, and a dust mask or respirator to prevent inhaling dust and fumes.

By having these must-have tools in your plumbing toolkit, you'll be well-equipped to handle a wide range of DIY plumbing tasks, from fixing leaky faucets to replacing sink traps. In the next section, we'll discuss the various materials and supplies you'll

Choosing the Right Pipes, Fittings, and Sealants

When it comes to DIY plumbing projects, selecting the appropriate pipes, fittings, and sealants is just as important as having the right tools. Using the correct materials ensures that your repairs and installations will be durable, leak-free, and compliant with local building codes. In this section, we'll discuss the various types of pipes, fittings, and sealants available, and provide guidance on how to choose the best options for your specific plumbing needs.

Pipes:
There are several types of pipes used in residential plumbing, each with its own advantages and best uses.

1. Copper Pipes:
Copper pipes are a popular choice for water supply lines due to their durability, resistance to corrosion, and ability to withstand high temperatures and pressures. They are available in rigid and flexible forms, and can be connected using soldered, compression, or push-fit fittings. Copper pipes are suitable for both hot and cold water applications, and have a proven track record of long-term reliability.

2. PEX (Cross-Linked Polyethylene) Pipes:
PEX pipes are a flexible, plastic alternative to copper pipes. They are lightweight, easy to install, and resistant to freezing and bursting. PEX pipes are connected using crimp, clamp, or push-fit fittings, and can be used for both hot and cold water supply lines. They are also less expensive than copper pipes, making them an attractive option for many homeowners.

3. PVC (Polyvinyl Chloride) Pipes:

PVC pipes are rigid, plastic pipes commonly used for drain, waste, and vent (DWV) lines. They are lightweight, easy to work with, and resistant to corrosion and chemical damage. PVC pipes are connected using solvent-welded fittings, which create strong, leak-proof joints. However, PVC pipes are not suitable for hot water applications, as they can warp and become brittle at high temperatures.

4. ABS (Acrylonitrile Butadiene Styrene) Pipes:

ABS pipes are another type of rigid, plastic pipe used for DWV lines. They are similar to PVC pipes in terms of ease of use and resistance to corrosion, but have a higher impact strength and are more resistant to cold temperatures. ABS pipes are also connected using solvent-welded fittings, and are not suitable for hot water applications.

Fittings:

Fittings are used to connect pipes, change direction, or branch out to multiple fixtures. The type of fitting you choose will depend on the type of pipe you are using and the specific connection you need to make.

1. Copper Fittings:

Copper fittings are used to connect copper pipes and are available in a variety of shapes, including elbows, tees, couplings, and adapters. These fittings can be soldered, compression-fitted, or push-fitted, depending on the application and the type of connection required.

2. PEX Fittings:

PEX fittings are designed specifically for use with PEX pipes and are available in crimp, clamp, or push-fit styles. These fittings create secure, leak-free connections and are easy to install without the need for soldering.

3. PVC and ABS Fittings:

PVC and ABS fittings are used to connect PVC and ABS pipes, respectively. These fittings are solvent-welded to the pipes using a special adhesive that creates a permanent, watertight bond. PVC and ABS fittings are available in a wide range of shapes and sizes to accommodate various DWV configurations.

Sealants:

Sealants are used to create watertight and airtight seals between pipes, fittings, and fixtures. They help prevent leaks and ensure the long-term durability of your plumbing system.

1. Teflon Tape (PTFE Tape):

Teflon tape, also known as plumber's tape, is a thin, white tape that is wrapped around the threads of pipes and fittings to create a watertight seal. It is commonly used on threaded connections, such as those found on supply lines and showerheads. Teflon tape is easy to apply and helps prevent leaks and thread corrosion.

2. Pipe Dope (Thread Sealant):

Pipe dope is a paste-like compound that is applied to the threads of pipes and fittings to lubricate and seal the connection. It helps prevent leaks and makes it easier to tighten and disassemble threaded connections. Pipe dope is available in both standard and lead-free formulations, with lead-free options being required for potable water applications.

3. Silicone Caulk:

Silicone caulk is a flexible, waterproof sealant that is used to fill gaps and create watertight seals around sinks, tubs, and other fixtures. It is easy to apply and remains flexible over time, allowing it to accommodate minor shifts and vibrations in the plumbing system. Silicone caulk is available in a variety of colors to match different fixture finishes.

When choosing pipes, fittings, and sealants for your DIY plumbing projects, always consider factors such as local building codes, the intended application (e.g., water supply lines, DWV lines), and compatibility with existing plumbing materials. By selecting the appropriate materials and using them correctly, you can ensure that your plumbing repairs and installations will stand the test of time.

Safety Equipment

When working on any DIY plumbing project, safety should always be your top priority. Plumbing work can involve a variety of potential hazards, including sharp edges, hot surfaces, chemicals, and heavy materials. By using the appropriate safety equipment and following key precautions, you can protect yourself from injuries and ensure a successful, incident-free plumbing project. In this section, we'll discuss the essential safety equipment you should use and the precautions you should take when working on plumbing tasks.

1. Safety Glasses or Goggles:
Protecting your eyes is crucial when working on plumbing projects. Safety glasses or goggles shield your eyes from debris, splashes, and other hazards that can cause eye injuries. Look for glasses or goggles that are impact-resistant, provide adequate coverage, and meet ANSI Z87.1 safety standards. Always wear eye protection when cutting, drilling, soldering, or handling chemicals.

2. Work Gloves:
Wear durable work gloves to protect your hands from sharp edges, hot surfaces, chemicals, and other potential hazards. Choose gloves that fit well and provide adequate dexterity for handling tools and materials. Leather, canvas, or synthetic gloves with reinforced palms and fingers are suitable for most plumbing tasks. When working with chemicals, use gloves made of neoprene, nitrile, or other chemical-resistant materials.

3. Dust Mask or Respirator:
When cutting, drilling, or sanding materials like PVC or ABS pipes, you may be exposed to dust and debris that can irritate your lungs. Use a dust mask or respirator to prevent inhaling these particles.

For tasks involving chemicals or strong odors, such as soldering or applying solvent-based adhesives, use a respirator with the appropriate cartridges to filter out harmful fumes.

4. Knee Pads:

Plumbing work often involves kneeling on hard surfaces for extended periods, which can cause discomfort and knee injuries. Wear knee pads to cushion your knees and reduce the stress on your joints. Look for knee pads with a durable outer shell and a comfortable, moisture-wicking lining.

5. Slip-Resistant Shoes:

Wear slip-resistant shoes or boots with good traction to prevent slips and falls, especially when working on wet surfaces or in areas with potential water spillage. Choose shoes with closed toes and sturdy soles to protect your feet from falling objects and provide support when lifting heavy materials.

Precautions:

1. Shut Off Water Supply:

Before starting any plumbing repair or installation, always shut off the water supply to the affected area. Locate the nearest shut-off valve, which may be a fixture shut-off valve or the main shut-off valve for your home, and turn it clockwise to close. This will prevent water from flowing through the pipes while you work, reducing the risk of leaks, floods, and water damage.

2. Drain Pipes:

After shutting off the water supply, open the faucets in the affected area to drain any remaining water from the pipes. This will relieve pressure in the system and make it easier to work on the pipes without the risk of water spraying or flooding.

3. Use Proper Ventilation:
When working with chemicals, such as solvent-based adhesives or cleaning products, ensure that your work area has adequate ventilation. Open windows and doors, and use fans to circulate fresh air and remove harmful fumes. If possible, work outdoors or in a well-ventilated garage or workshop.

4. Handle Tools and Materials Safely:
Always use tools and materials according to the manufacturer's instructions and safety guidelines. Keep tools sharp, clean, and in good repair to ensure they work properly and reduce the risk of accidents. When handling heavy materials, like pipes or fixtures, use proper lifting techniques to avoid back injuries. Bend at the knees, keep your back straight, and lift with your legs, not your back.

5. Keep Your Work Area Clean and Organized:
Maintain a clean, organized work area to reduce the risk of trips, falls, and other accidents. Keep tools and materials neatly arranged and put away when not in use. Clean up any spills or debris promptly, and dispose of waste materials properly.

6. Be Mindful of Electrical Hazards:
Plumbing work often takes place near electrical outlets, switches, and appliances. Be cautious when working around electricity, and avoid contact with water if there are any exposed electrical components nearby. If you are unsure about the safety of working near electrical hazards, consult a licensed electrician before proceeding.

7. Know Your Limits:

DIY plumbing projects can be rewarding and cost-effective, but it's essential to know your limits and when to call in a professional. If you encounter a plumbing issue that is beyond your skill level or involves major repairs or modifications, don't hesitate to contact a licensed plumber. Attempting to tackle complex plumbing problems without the necessary knowledge and experience can lead to costly mistakes and potential safety hazards.

By using the appropriate safety equipment and following these precautions, you can minimize the risks associated with DIY plumbing work and ensure a safe, successful project. Remember, investing in your safety is always worth the effort, as it can help prevent injuries and protect your well-being while you tackle your plumbing tasks.

Chapter 3
Fixing Common Toilet Issues
Diagnosing and Repairing a Running Toilet

Toilets are one of the most essential fixtures in any home, and when they malfunction, it can lead to frustration, water waste, and potential damage to your property. One of the most common toilet issues homeowners face is a running toilet, which can cause a constant flow of water into the bowl and result in higher water bills. In this section, we'll discuss how to diagnose and repair a running toilet, so you can restore your toilet's proper function and save water in the process.

Diagnosing and Repairing a Running Toilet:

Step 1: Identify the Cause
A running toilet can be caused by several factors, including a faulty flapper, a misadjusted float, or a defective fill valve. To determine the cause of your running toilet, start by removing the tank lid and observing the components inside.

Step 2: Check the Flapper
The flapper is the rubber seal that covers the flush valve opening at the bottom of the tank. When you flush the toilet, the flapper lifts to allow water to flow from the tank into the bowl, and then settles back down to seal the opening. If the flapper is worn, damaged, or misaligned, it may not seal properly, causing water to continually leak into the bowl.

To check the flapper:
a. Flush the toilet and observe the flapper's movement. It should lift completely and then settle back down firmly over the flush valve opening.

29

b. If the flapper appears worn or damaged, or if it doesn't seal properly, it needs to be replaced.

c. To replace the flapper, turn off the water supply to the toilet, flush to drain the tank, and then disconnect the old flapper from the flush lever arm and the flush valve. Install a new flapper of the same size and type, and reconnect it to the flush lever arm and flush valve.

Step 3: Adjust the Float

The float is the device that signals the fill valve to shut off when the tank has reached the appropriate water level. If the float is set too high, the water level in the tank will be too high, causing water to continually flow into the overflow tube and down into the bowl.

To adjust the float:

a. Locate the float, which may be a ball float (attached to a rod) or a cup float (attached to the fill valve).

b. For a ball float, locate the screw or clip that connects the float rod to the fill valve. Turn the screw clockwise to lower the float or counterclockwise to raise it. Adjust the float so that the water level in the tank is about 1/2 inch below the top of the overflow tube.

c. For a cup float, pinch the float clip and slide the float up or down along the fill valve shaft to adjust the water level. The water should stop about 1/2 inch below the top of the overflow tube.

Step 4: Inspect the Fill Valve

The fill valve controls the flow of water into the tank and is responsible for shutting off the water when the tank is full. If the fill valve is faulty, it may not shut off properly, causing water to continually run into the tank and down the overflow tube.

To inspect the fill valve:

a. Flush the toilet and observe the fill valve. It should shut off completely when the tank is full.

b. If the fill valve continues to run or if it makes a hissing or chattering noise, it may need to be cleaned or replaced.

c. To clean the fill valve, turn off the water supply and remove the cap on top of the valve. Flush the valve by holding a cup over the uncapped valve and turning the water supply back on briefly. This will dislodge any debris that may be blocking the valve. Replace the cap and turn the water supply back on.

d. If cleaning the fill valve doesn't solve the problem, you may need to replace it. To do this, turn off the water supply, disconnect the supply line, and remove the old fill valve. Install a new fill valve according to the manufacturer's instructions, and reconnect the supply line.

Step 5: Test and Monitor

After making any repairs or adjustments, flush the toilet several times to ensure that it is functioning properly and that the running has stopped. Monitor the toilet over the next few days to make sure the issue has been resolved.

By following these steps and addressing the most common causes of a running toilet, you can often resolve the issue yourself without the need for a professional plumber. However, if you're unsure about any part of the process or if the problem persists after attempting repairs, don't hesitate to contact a licensed plumber for assistance.

Regular maintenance, such as cleaning the components inside the tank and replacing worn parts as needed, can help prevent running toilets and other issues from occurring. By taking a proactive approach to toilet maintenance, you can ensure that your toilet continues to function efficiently and effectively, saving you water and money in the long run.

Unclogging a Toilet Like a Pro

A clogged toilet is a common household problem that can cause stress and frustration, especially if you're not sure how to tackle the issue effectively. Knowing the right techniques and tools to use can help you unclog your toilet quickly and efficiently, saving you time and potential embarrassment. In this section, we'll discuss professional methods for unclogging a toilet, so you can handle this plumbing issue with confidence.

Step 1: Assess the Situation

Before attempting to unclog the toilet, take a moment to assess the situation. If the water level in the bowl is high and close to overflowing, it's best to wait 10-15 minutes to allow the water to recede naturally. This will minimize the risk of overflow and make it easier to work on the clog. If the water level is normal or low, you can proceed with unclogging the toilet.

Step 2: Use a Plunger

A plunger is the most common tool for unclogging toilets and is often the first line of defense against clogs. To use a plunger effectively:

a. Choose the right plunger: For toilets, use a flange plunger, which has a soft rubber flap extending from the bottom of the cup. This flap fits snugly into the toilet bowl drain, creating a better seal for plunging.

b. Ensure there is enough water in the bowl to cover the plunger head. If the water level is low, add some water from a bucket or cup until the plunger head is submerged.

c. Place the plunger over the toilet bowl drain, ensuring that the flange is inserted into the drain opening.

d. Begin plunging vigorously, maintaining the seal between the plunger and the drain. Plunge straight up and down, using quick, powerful strokes. Continue plunging for 20-30 seconds.

e. After plunging, remove the plunger and observe the water level in the bowl. If the water recedes quickly, the clog has likely been dislodged. Flush the toilet to confirm that it is draining properly.

f. If the water level remains high or drains slowly, repeat the plunging process. You may need to plunge several times to fully remove the clog.

Step 3: Use a Toilet Auger (Plumbing Snake)
If plunging doesn't dislodge the clog, the next step is to use a toilet auger, also known as a plumbing snake. A toilet auger is a specialized tool designed to reach deep into the toilet trap and break up stubborn clogs. To use a toilet auger:

a. Insert the auger head into the toilet bowl, guiding it into the drain opening.

b. Crank the handle clockwise to extend the auger cable into the drain. Continue cranking until you feel resistance, indicating that you've reached the clog.

c. Once you've reached the clog, continue cranking the handle to break up the blockage. You may need to crank back and forth several times to fully dislodge the clog.

d. After breaking up the clog, retract the auger cable by cranking the handle counterclockwise. Remove the auger from the toilet bowl.

e. Flush the toilet to check if the clog has been removed and the water is draining properly. If the toilet is still clogged, repeat the process with the auger.

Step 4: Try Chemical Drain Cleaners (Last Resort)

If plunging and using a toilet auger have not successfully removed the clog, you may consider using a chemical drain cleaner as a last resort. However, it's essential to use these products cautiously, as they can be harmful to your plumbing and the environment if used improperly.

a. Choose a drain cleaner specifically designed for toilets, and follow the manufacturer's instructions carefully.

b. Pour the recommended amount of drain cleaner into the toilet bowl, and allow it to sit for the specified time (usually 15-30 minutes).

c. After the waiting period, flush the toilet to see if the clog has been removed. If the toilet is still clogged, do not add more drain cleaner, as this can damage your plumbing.

d. If the chemical drain cleaner does not work, it's best to contact a professional plumber to address the issue.

Step 5: Prevent Future Clogs

To minimize the risk of future toilet clogs, adopt these preventive measures:

a. Only flush human waste and toilet paper down the toilet. Avoid flushing items like wipes, cotton swabs, feminine hygiene products, or other debris that can cause clogs.

b. Use toilet paper conservatively, and avoid using excessive amounts in a single flush.

c. Encourage family members and guests to follow proper toilet use guidelines to prevent clogs and plumbing issues.

d. Consider installing a non-clogging toilet or a pressure-assisted toilet, which are designed to handle larger waste loads and reduce the risk of clogs.

By following these steps and using the right tools and techniques, you can effectively unclog your toilet like a professional. However, if you encounter a particularly stubborn clog or if you're unsure about any part of the process, don't hesitate to contact a licensed plumber for assistance. Regular toilet maintenance and proper use can go a long way in preventing clogs and keeping your plumbing system functioning smoothly.

Replacing a Toilet Flapper or Fill Valve

Two of the most common components that may need replacing in a toilet are the flapper and the fill valve. The flapper is responsible for sealing the flush valve and controlling the release of water from the tank into the bowl, while the fill valve regulates the water level in the tank and refills it after each flush. Over time, these parts can wear out, causing issues like a running toilet or inadequate flushing. In this section, we'll discuss how to replace a toilet flapper and fill valve, so you can maintain your toilet's proper function and efficiency.

Replacing a Toilet Flapper:

Step 1: Identify the Problem
A malfunctioning flapper can cause a running toilet, as it may not seal properly, allowing water to continually leak from the tank into the bowl. To confirm that the flapper needs replacement, perform a simple dye test:

a. Remove the tank lid and add a few drops of food coloring or dye to the water in the tank.

b. Wait 15-20 minutes without flushing the toilet.

c. If the colored water appears in the bowl, the flapper is likely leaking and needs to be replaced.

Step 2: Choose the Right Replacement Flapper
Flappers come in various sizes and styles to fit different toilet models. To ensure you purchase the correct replacement flapper:

a. Note the brand and model of your toilet, which is usually printed on the inside of the tank or stamped on the underside of the tank lid.

b. Take the old flapper to a hardware store or home improvement center to find an exact match or a compatible universal flapper.

Step 3: Install the New Flapper

a. Turn off the water supply to the toilet by closing the shutoff valve, typically located near the base of the toilet.

b. Flush the toilet to drain the tank, and sponge out any remaining water.

c. Disconnect the old flapper from the flush lever chain and the flush valve.

d. Clean the flush valve seat to ensure a proper seal with the new flapper.

e. Connect the new flapper to the flush valve, ensuring it is properly aligned and seated.

f. Attach the new flapper to the flush lever chain, adjusting the chain length so that the flapper opens fully when the toilet is flushed but has slight slack when closed.

g. Turn the water supply back on and flush the toilet to test the new flapper. Make any necessary adjustments to the chain length for optimal performance.

Replacing a Fill Valve:

Step 1: Identify the Problem
A faulty fill valve can cause a running toilet, inadequate tank filling, or a toilet that shuts off slowly. If you've ruled out a malfunctioning flapper and the fill valve appears to be the issue, it's time to replace it.

Step 2: Choose the Right Replacement Fill Valve

Fill valves come in various designs, including ball-cock, diaphragm, and float-cup types. To select the correct replacement fill valve:

a. Note the brand and model of your toilet, as some manufacturers require specific fill valve types.

b. Measure the height of your current fill valve from the base of the tank to the top of the valve to ensure the new valve will fit properly.

c. Choose a fill valve with adjustable height settings to accommodate your toilet's specifications.

Step 3: Install the New Fill Valve

a. Turn off the water supply to the toilet and flush to drain the tank.

b. Disconnect the water supply line from the old fill valve.

c. Remove the old fill valve by unscrewing the lock nut at the base of the valve underneath the tank.

d. Clean the hole in the tank where the fill valve sits to ensure a proper seal.

e. Insert the new fill valve through the hole and secure it with the provided lock nut.

f. Connect the water supply line to the new fill valve.

g. Adjust the height of the fill valve according to the manufacturer's instructions, ensuring that the water level in the tank is about 1/2 inch below the top of the overflow tube.

h. Turn the water supply back on and flush the toilet to test the new fill valve. Make any necessary adjustments to the water level for optimal performance.

By following these steps and choosing the appropriate replacement parts, you can successfully replace a toilet flapper or fill valve, ensuring that your toilet continues to function efficiently. Regular maintenance and inspection of these components can help you identify potential issues early, preventing more severe problems and costly repairs down the line.

If you're unsure about any part of the replacement process or encounter complications, don't hesitate to consult a professional plumber for guidance or assistance. With the right knowledge and tools, however, replacing a toilet flapper or fill valve is a manageable DIY task that can save you time and money while keeping your toilet in top working condition.

Chapter 4
Solving Sink and Faucet Problems
Fixing Leaky Faucets and Handles

Leaky faucets and handles are common plumbing issues that can waste water, create annoying drips, and lead to more severe problems if left unaddressed. Fortunately, fixing these issues is often a straightforward DIY task that can be accomplished with basic tools and a little know-how. In this section, we'll discuss how to identify and fix leaky faucets and handles, so you can keep your sinks and faucets functioning properly and efficiently.

Fixing Leaky Faucets and Handles:

Step 1: Identify the Type of Faucet
Before attempting to fix a leaky faucet or handle, it's essential to identify the type of faucet you have, as the repair process can vary depending on the design. The four most common types of faucets are:

a. Compression Faucets: These faucets have separate handles for hot and cold water and use rubber washers to control water flow.

b. Cartridge Faucets: These faucets have a single handle that controls both temperature and water flow using a cartridge inside the faucet body.

c. Ceramic Disk Faucets: These faucets also have a single handle but use a ceramic disk valve to control water flow and temperature.

d. Ball Faucets: These faucets have a single handle that controls water flow and temperature using a ball-shaped valve with slots and springs.

41

Step 2: Gather Tools and Materials

Before starting the repair, gather the necessary tools and materials, which may include:

a. Adjustable wrench or pliers

b. Flathead and Phillips screwdrivers

c. Replacement parts (washers, O-rings, cartridges, or ceramic disks, depending on the faucet type)

d. Plumber's grease

e. Cloth or towel

Step 3: Turn Off the Water Supply

Locate the shut-off valves beneath the sink and turn them clockwise to shut off the water supply to the faucet. If there are no shut-off valves, turn off the main water supply to the house.

Step 4: Disassemble the Faucet

a. For Compression Faucets:
- Remove the handle screw and handle
- Unscrew the nut holding the stem in place
- Remove the stem and inspect the rubber washer at the bottom
- Replace the washer if it's worn or damaged

b. For Cartridge Faucets:
- Remove the handle screw and handle
- Remove the retaining clip or nut holding the cartridge in place
- Pull out the cartridge and inspect it for damage
- Replace the cartridge if necessary

c. For Ceramic Disk Faucets:
- Remove the handle screw and handle
- Remove the escutcheon cap and disk cylinder
- Inspect the ceramic disk for cracks or damage
- Replace the ceramic disk if necessary

d. For Ball Faucets:

- Remove the handle screw and handle
- Remove the cap and collar
- Lift out the ball valve and inspect the rubber seats and springs
- Replace the seats and springs if they're worn or damaged

Step 5: Reassemble the Faucet

a. Apply a small amount of plumber's grease to the new parts (washers, O-rings, cartridges, or ceramic disks) before installation.

b. Reassemble the faucet in the reverse order of disassembly, making sure all parts are properly aligned and secured.

c. Turn the water supply back on and test the faucet for leaks and proper function.

Step 6: Troubleshoot Additional Issues

If the faucet continues to leak after replacing the necessary parts, check for the following issues:

a. Ensure all connections are tight and properly sealed.

b. Inspect the faucet body for cracks or damage, which may require replacing the entire faucet.

c. Check for leaks around the base of the faucet, which may indicate a worn-out O-ring or a loose mounting nut.

By following these steps and identifying the specific type of faucet you have, you can successfully fix most leaky faucets and handles. Regular maintenance, such as cleaning the aerator and inspecting for leaks, can help prevent more serious issues from developing and prolong the life of your faucets.

If you encounter a particularly challenging leak or are unsure about any part of the repair process, don't hesitate to contact a professional plumber for assistance. With the right tools, parts, and knowledge, however, fixing leaky faucets and handles is a manageable DIY task that can save you money and keep your sinks functioning properly.

Unclogging Sink Drains with Natural and Chemical Solutions

Clogged sink drains are a common household problem that can cause slow drainage, unpleasant odors, and even water backup. While there are many commercial drain cleaners available, some homeowners prefer to use natural or less harsh chemical solutions to clear their drains. In this section, we'll discuss both natural and chemical methods for unclogging sink drains, so you can choose the approach that best suits your needs and preferences.

Natural Solutions:

1. Boiling Water
One of the simplest and most effective methods for clearing minor clogs is to pour boiling water down the drain.

a. Boil a pot of water and carefully pour it directly into the drain opening.
b. Allow the hot water to work through the clog for a few minutes.
c. Repeat the process several times to break up and flush away the clog.

2. Baking Soda and Vinegar
The combination of baking soda and vinegar creates a fizzing reaction that can help dissolve and break up clogs.

a. Remove standing water from the sink using a cup or bowl.
b. Pour 1/2 cup of baking soda down the drain.
c. Follow with 1/2 cup of white vinegar.
d. Cover the drain with a stopper or cloth to contain the fizzing reaction.
e. Let the mixture sit for 15-30 minutes.
f. Flush the drain with hot water to rinse away the clog and the baking soda and vinegar mixture.

3. Salt, Borax, and Vinegar

For tougher clogs, a combination of salt, borax, and vinegar can be more effective.

a. Mix 1/4 cup of salt, 1/4 cup of Borax, and 1/2 cup of vinegar in a bowl.
b. Pour the mixture down the drain and let it sit for 15-30 minutes.
c. Flush the drain with boiling water to clear the clog and rinse away the cleaning mixture.

Chemical Solutions:

1. Plunging with Drain Cleaner

For more stubborn clogs, using a plunger in combination with a chemical drain cleaner can be effective.

a. Choose a drain cleaner suitable for your type of sink and follow the manufacturer's instructions carefully.
b. Pour the recommended amount of drain cleaner into the sink and let it sit for the specified time.
c. While the cleaner is working, fill the sink with enough hot water to cover the plunger head.
d. Place the plunger over the drain opening, ensuring a good seal.
e. Plunge vigorously for 20-30 seconds, using quick, powerful strokes.
f. Rinse the drain with hot water to flush away the clog and the drain cleaner.

2. Enzyme-Based Drain Cleaners

Enzyme-based drain cleaners contain bacteria that feed on organic materials, such as hair and food particles, breaking them down over time.

a. Choose an enzyme-based drain cleaner and follow the manufacturer's instructions.

b. Pour the specified amount of cleaner into the drain and let it sit for the recommended time, usually several hours or overnight.

c. Flush the drain with hot water to rinse away the broken-down clog material.

Safety Precautions and Tips:

- Always wear gloves and protective eyewear when using chemical drain cleaners.
- Avoid using chemical drain cleaners too frequently, as they can damage pipes over time.
- Do not mix different types of drain cleaners, as this can cause dangerous chemical reactions.
- If natural methods and plunging do not clear the clog, it's best to contact a professional plumber to avoid damaging your plumbing system.

By trying these natural and chemical solutions for unclogging sink drains, you can often resolve minor to moderate clogs without the need for professional assistance. However, if you encounter a persistent or severe clog, or if you're unsure about any part of the cleaning process, it's always best to consult a licensed plumber for guidance and assistance.

Regular maintenance, such as using strainers to catch debris and periodically flushing drains with hot water or a mixture of baking soda and vinegar, can help prevent clogs from forming and keep your sink drains flowing freely.

Installing a New Faucet or Sink Strainer

Updating your sink faucet or strainer can improve both the functionality and the aesthetic appeal of your kitchen or bathroom. Whether you're replacing a worn-out faucet or upgrading to a new style, installing a new faucet or sink strainer is a manageable DIY project for most homeowners. In this section, we'll provide step-by-step instructions for installing a new faucet and sink strainer, so you can tackle this task with confidence.

Installing a New Faucet:

Step 1: Choose the Right Faucet
Before purchasing a new faucet, consider the following factors:

a. Sink Compatibility: Ensure that the new faucet is compatible with your sink's configuration (e.g., number of holes, hole spacing).
b. Faucet Type: Decide on the type of faucet you want, such as a single-handle, double-handle, or touchless model.
c. Finish and Style: Choose a finish and style that complements your sink and overall décor.

Step 2: Gather Tools and Materials
You'll need the following tools and materials to install your new faucet:

a. Adjustable wrench or basin wrench
b. Plumber's putty or silicone sealant
c. Flexible supply lines
d. Thread seal tape (Teflon tape)

Step 3: Remove the Old Faucet

a. Turn off the water supply to the faucet by closing the shut-off valves beneath the sink.
b. Disconnect the supply lines from the faucet using an adjustable wrench.
c. Remove the mounting nuts that secure the faucet to the sink using a basin wrench.
d. Lift the old faucet out of the sink and clean the sink surface thoroughly.

Step 4: Install the New Faucet

a. Place the faucet gasket or deck plate (if applicable) over the sink holes.
b. Apply a bead of plumber's putty or silicone sealant around the base of the faucet.
c. Insert the faucet through the sink holes and secure it with the provided mounting nuts, tightening them with a basin wrench.
d. Connect the flexible supply lines to the faucet and the shut-off valves, using thread seal tape on the threaded connections.
e. Turn on the water supply and check for leaks, tightening connections if necessary.
f. Run the faucet to flush out any debris and ensure proper function.

Installing a New Sink Strainer:

Step 1: Choose the Right Strainer
Sink strainers come in various finishes and styles to match your sink and faucet. Ensure that the strainer you choose is compatible with your sink's drain opening size.

Step 2: Gather Tools and Materials

You'll need the following tools and materials to install your new sink strainer:

a. Adjustable wrench or pliers
b. Plumber's putty
c. Tailpiece and slip nut (if not included with the strainer)

Step 3: Remove the Old Strainer

a. Disconnect the drain tailpiece from the strainer by loosening the slip nut with an adjustable wrench.
b. Remove the old strainer by unscrewing the locknut from beneath the sink.
c. Clean the sink surface around the drain opening, removing any old putty or debris.

Step 4: Install the New Strainer

a. Roll a small bead of plumber's putty between your hands and apply it around the underside of the strainer's lip.
b. Insert the strainer into the sink's drain opening, pressing it firmly into place.
c. From beneath the sink, place the rubber gasket, friction ring, and locknut onto the strainer's threaded shaft.
d. Tighten the locknut by hand, then give it a quarter turn with an adjustable wrench to secure it in place.
e. Remove any excess putty from around the strainer in the sink basin.
f. Connect the drain tailpiece to the strainer using the slip nut, tightening it with an adjustable wrench.
g. Run water through the strainer to check for leaks and ensure proper drainage.

By following these step-by-step instructions, you can successfully install a new faucet or sink strainer, updating the look and functionality of your sink. Remember to take your time, double-check all connections, and always turn off the water supply before beginning any plumbing project.

If you encounter any challenges during the installation process or are unsure about any aspect of the project, don't hesitate to consult a professional plumber for guidance or assistance. With the right tools, materials, and knowledge, however, installing a new faucet or sink strainer is a rewarding DIY task that can save you money and give your sink a fresh, updated look.

Chapter 5
Tackling Shower and Bathtub Issues
Fixing Low Water Pressure in Your Shower

Low water pressure in your shower can be frustrating and can make your bathing experience less enjoyable. There are several potential causes for low shower water pressure, ranging from simple clogs to more complex plumbing issues. In this section, we'll discuss how to diagnose and fix low water pressure in your shower, so you can restore your shower's performance and enjoy a satisfying bathing experience.

Step 1: Identify the Scope of the Problem
First, determine whether the low water pressure is specific to your shower or if it affects other fixtures in your home.

a. Check other faucets and showers to see if they also have low water pressure.
b. If the problem is isolated to your shower, the issue is likely with the showerhead or the shower's plumbing.
c. If the problem affects multiple fixtures, the issue may be with your home's main water supply or pressure regulator.

Step 2: Clean the Showerhead
Mineral buildup and sediment can clog showerhead nozzles, reducing water flow and pressure. To clean your showerhead:

a. Remove the showerhead by unscrewing it counterclockwise from the shower arm.
b. Soak the showerhead in a mixture of equal parts white vinegar and water for several hours or overnight.
c. Scrub the nozzles with an old toothbrush to remove any remaining buildup.

d. Rinse the showerhead thoroughly and reattach it to the shower arm.

e. Run the shower to check for improved water pressure.

Step 3: Check for Leaks

Leaks in the shower's plumbing can divert water and reduce pressure at the showerhead. To check for leaks:

a. Inspect the shower arm and the connection between the showerhead and arm for visible leaks or drips.

b. Check the shower valve for leaks by removing the trim plate and looking for signs of water damage or dripping.

c. If leaks are present, tighten connections or replace damaged components as necessary.

Step 4: Adjust the Shower Valve

Some shower valves have built-in pressure regulators or flow restrictors that can limit water pressure. To adjust the valve:

a. Remove the shower valve's trim plate and handle to access the valve stem.

b. Look for a small screw or knob on the valve stem that adjusts the pressure or flow.

c. Turn the screw or knob slightly counterclockwise to increase water pressure, being careful not to overtighten.

d. Replace the handle and trim plate, and test the shower for improved pressure.

Step 5: Check the Main Water Supply and Pressure Regulator

If the low water pressure affects multiple fixtures in your home, the issue may be with your main water supply or pressure regulator.

a. Locate your home's main water shut-off valve and pressure regulator (usually near the water meter or where the main water line enters your home).
b. Check the pressure regulator for proper function, adjusting it if necessary according to the manufacturer's instructions.
c. If the main water shut-off valve is not fully open, turn it counterclockwise to open it completely.
d. If adjusting the pressure regulator or opening the main shut-off valve does not improve water pressure, contact your local water utility company to inquire about possible issues with the municipal water supply.

Step 6: Consider Replacing the Showerhead
If cleaning and adjusting your current showerhead does not improve water pressure, consider replacing it with a new, high-pressure showerhead designed to maximize flow and pressure.

a. Choose a showerhead with a higher flow rate (measured in gallons per minute, or GPM) and pressure-enhancing features, such as self-pressurizing nozzles or air-induction technology.
b. Ensure that the new showerhead is compatible with your shower arm and plumbing.
c. Install the new showerhead according to the manufacturer's instructions and test for improved water pressure.

By following these steps and troubleshooting the various potential causes of low shower water pressure, you can often resolve the issue and restore your shower's performance. However, if you encounter a persistent problem or are unsure about any aspect of the diagnostic or repair process, it's always best to consult a licensed plumber for professional assistance.

Regular maintenance, such as cleaning your showerhead and checking for leaks, can help prevent low water pressure issues from developing and ensure that your shower continues to function efficiently. By addressing low water pressure promptly and effectively, you can maintain a comfortable and enjoyable showering experience in your home.

Replacing a Shower Head or Bathtub Spout

Updating your shower head or bathtub spout can improve the functionality, appearance, and water efficiency of your bathroom. Whether you're looking to replace a leaky, outdated, or low-performing fixture, installing a new shower head or bathtub spout is a relatively simple DIY project. In this section, we'll provide step-by-step instructions for replacing both shower heads and bathtub spouts.

Replacing a Shower Head:

Step 1: Choose a new shower head
a. Consider factors such as water efficiency, spray pattern options, and style when selecting a new shower head.
b. Ensure that the new shower head is compatible with your existing shower arm and plumbing.

Step 2: Remove the old shower head
a. Use an adjustable wrench to loosen the connection between the shower head and the shower arm.
b. Turn the shower head counterclockwise to unscrew it from the shower arm.
c. If the shower head is difficult to remove, apply penetrating oil to the connection and let it sit for a few minutes before trying again.

Step 3: Clean the shower arm threads
a. Use a wire brush or an old toothbrush to remove any rust, mineral buildup, or old plumber's tape from the threads of the shower arm.
b. Ensure that the threads are clean and smooth to create a leak-free connection with the new shower head.

Step 4: Apply plumber's tape

a. Wrap plumber's tape (also known as Teflon tape) around the threads of the shower arm in a clockwise direction.

b. Overlap the tape by about 50% as you wrap it around the threads 2-3 times.

c. The plumber's tape helps create a watertight seal and prevents leaks.

Step 5: Install the new shower head

a. Screw the new shower head onto the shower arm by turning it clockwise.

b. Tighten the connection using your hands, ensuring that the shower head is securely in place.

c. Use an adjustable wrench to give the shower head a quarter-turn, being careful not to overtighten, as this may damage the shower head or cause leaks.

Step 6: Test the new shower head

a. Turn on the water supply and test the new shower head for proper function and any leaks.

b. If leaks are present, double-check that the connections are tight and that the plumber's tape is properly applied.

Replacing a Bathtub Spout:

Step 1: Determine the type of bathtub spout

a. Slip-on spouts: These spouts simply slip onto the pipe coming out of the wall and are held in place by an Allen screw.

b. Threaded spouts: These spouts screw onto the threaded pipe coming out of the wall.

Step 2: Remove the old bathtub spout

a. For slip-on spouts:

- Locate the Allen screw underneath or on the back of the spout.

- Use an Allen wrench to loosen the screw and remove the spout by pulling it straight off the pipe.

b. For threaded spouts:
- Turn the spout counterclockwise to unscrew it from the pipe.
- If the spout is difficult to remove, apply penetrating oil to the threads and let it sit for a few minutes before trying again.

Step 3: Clean the pipe and threads

a. Use a wire brush or an old toothbrush to remove any debris, mineral buildup, or old plumber's tape from the pipe or threads.

b. Ensure that the surface is clean and smooth to create a proper seal with the new spout.

Step 4: Install the new bathtub spout

a. For slip-on spouts:
- Slide the new spout onto the pipe, ensuring that it is fully seated against the wall.
- Secure the spout in place by tightening the Allen screw with an Allen wrench.

b. For threaded spouts:
- Wrap plumber's tape around the threads of the pipe in a clockwise direction, overlapping the tape by about 50% and making 2-3 wraps.
- Screw the new spout onto the threaded pipe by turning it clockwise, tightening it by hand.
- Use an adjustable wrench to give the spout a quarter-turn, being careful not to overtighten.

Step 5: Test the new bathtub spout

a. Turn on the water supply and test the new spout for proper function and any leaks.

b. If leaks are present, double-check that the connections are tight and that the plumber's tape (for threaded spouts) is properly applied.

By following these step-by-step instructions, you can successfully replace your shower head or bathtub spout, updating the look and performance of your bathroom fixtures. Remember to take your time, double-check all connections, and always turn off the water supply before beginning any plumbing project.

If you encounter any challenges during the replacement process or are unsure about any aspect of the project, don't hesitate to consult a professional plumber for guidance or assistance. With the right tools, materials, and knowledge, however, replacing a shower head or bathtub spout is a manageable DIY task that can save you money and give your bathroom a fresh, updated look.

Repairing Leaky Shower Valves or Diverters

A leaky shower valve or diverter can waste water, cause damage to your bathroom, and create an annoying dripping sound. In this section, we'll discuss how to diagnose and repair leaky shower valves or diverters, so you can stop leaks and restore your shower's proper function.

Diagnosing the Problem:

1. Identify the type of shower valve you have:

a. Single-handle valve: Controls both water temperature and flow with one handle.

b. Double-handle valve: Has separate handles for hot and cold water.

c. Three-handle valve: Has separate handles for hot and cold water, plus a central diverter handle to switch between the showerhead and tub spout.

2. Determine the source of the leak:

a. Showerhead: Water drips from the showerhead when the shower is turned off.

b. Tub spout: Water leaks from the tub spout when the shower is in use.

c. Shower valve: Water leaks from the valve handles or trim plate.

Repairing a Leaky Shower Valve:

Step 1: Turn off the water supply

Locate the shut-off valves for your shower (usually in the basement or crawl space) and turn them clockwise to shut off the water supply.

Step 2: Remove the shower valve trim and handle

a. Remove the decorative trim plate and handle from the shower valve.

b. If your valve has a screw-on escutcheon plate, unscrew it to access the valve.

Step 3: Identify the type of cartridge or stem

a. Single-handle valves use a cartridge, while double-handle and three-handle valves use stems.

b. Take a picture or make a note of the cartridge or stem type and orientation for reassembly.

Step 4: Remove the cartridge or stems

a. For cartridge valves, pull out the retaining clip and carefully remove the cartridge.

b. For stem valves, unscrew the bonnet nut and remove the stem.

Step 5: Replace O-rings and seals

a. Inspect the cartridge or stems for damage and replace if necessary.

b. Replace the O-rings and seals on the cartridge or stems with new ones, applying a small amount of plumber's grease.

Step 6: Reinstall the cartridge or stems

a. Insert the cartridge or stems back into the valve body, ensuring proper orientation.

b. Secure the cartridge with the retaining clip or screw the bonnet nut back onto the stem.

Step 7: Reassemble the trim and handle

a. Replace the escutcheon plate, trim plate, and handle.

b. Turn the water supply back on and test the shower for leaks.

Repairing a Leaky Diverter:

Step 1: Remove the diverter handle
a. Remove the screw that holds the diverter handle in place.
b. Pull off the handle to expose the diverter stem.

Step 2: Replace the diverter washer
a. Unscrew the diverter stem from the valve body.
b. Remove the old, worn-out rubber washer from the end of the diverter stem.
c. Replace the washer with a new one of the same size and thickness.

Step 3: Reassemble the diverter
a. Screw the diverter stem back into the valve body.
b. Replace the diverter handle and secure it with the screw.
c. Turn the water supply back on and test the diverter for leaks and proper function.

Tips for Preventing Leaks:

1. Replace washers, O-rings, and seals every few years to prevent wear and tear.
2. Avoid hanging items from the showerhead or diverter, as this can cause stress on the components and lead to leaks.
3. Be gentle when turning shower handles or diverters, as excessive force can damage the valve and cause leaks.

By following these steps and tips, you can effectively repair leaky shower valves or diverters and prevent future leaks. However, if you're unsure about any part of the repair process or encounter a more complex issue, it's always best to consult a professional plumber to ensure the problem is resolved correctly and safely.

Chapter 6
Maintaining and Repairing Water Heaters
Understanding the Different Types of Water Heaters

A water heater is an essential appliance in every home, providing hot water for bathing, cleaning, and other daily tasks. Proper maintenance and timely repairs can extend the life of your water heater and ensure a reliable supply of hot water. In this section, we'll discuss the different types of water heaters and their maintenance and repair requirements.

Understanding the Different Types of Water Heaters:

1. Conventional Storage Tank Water Heaters:
   a. These are the most common type of water heaters, consisting of an insulated tank that stores and heats water.
   b. They are available in electric, gas, and oil-fired models.
   c. Storage tank water heaters typically hold 20 to 80 gallons of water, depending on the model and household size.
   d. They continuously heat and store water, so hot water is always available when needed.

Maintenance and Repair:
- Flush the tank annually to remove sediment buildup, which can reduce efficiency and cause damage.
- Test the pressure relief valve regularly to ensure it's functioning properly.
- Check the anode rod every few years and replace it if it's badly corroded.
- If the water heater is leaking or not heating water properly, it may need repairs or replacement.

2. Tankless (On-Demand) Water Heaters:

a. These water heaters heat water instantly as it flows through the unit, rather than storing it in a tank.

b. They are available in electric and gas-fired models.

c. Tankless water heaters are more energy-efficient than storage tank models, as they only heat water when it's needed.

d. They have a longer lifespan than storage tank water heaters and take up less space.

Maintenance and Repair:
- Flush the unit annually to remove mineral buildup, which can reduce efficiency and cause damage.
- Clean or replace the water filter regularly to maintain water flow and pressure.
- If the water heater is not heating water properly or is displaying an error code, it may need repairs or replacement.

3. Heat Pump Water Heaters:

a. These water heaters use electricity to move heat from the air or ground into the water, rather than generating heat directly.

b. They are more energy-efficient than conventional electric water heaters, as they use less electricity to heat the water.

c. Heat pump water heaters typically have a higher upfront cost but can result in significant energy savings over time.

Maintenance and Repair:
- Clean the air filter regularly to maintain efficiency and prevent damage to the compressor.
- Flush the tank annually to remove sediment buildup.
- If the water heater is not heating water properly or is making unusual noises, it may need repairs or replacement.

4. Solar Water Heaters:

a. These water heaters use solar panels to collect heat from the sun and transfer it to the water.

b. They can be used in conjunction with a backup gas or electric water heater for cloudy days or high-demand periods.

c. Solar water heaters are environmentally friendly and can significantly reduce energy costs.

Maintenance and Repair:
- Check the solar panels regularly for damage or debris that may reduce efficiency.
- Flush the tank annually to remove sediment buildup.
- If the water heater is not heating water properly or is leaking, it may need repairs or replacement.

Regardless of the type of water heater you have, it's essential to schedule regular maintenance and address any issues promptly to prevent further damage and ensure a steady supply of hot water. Some common signs that your water heater may need repairs include:

- Insufficient hot water
- Water that's too hot
- Leaks or puddles around the base of the unit
- Strange noises coming from the water heater
- Discolored or smelly hot water

If you notice any of these issues or are unsure about how to maintain or repair your water heater, it's best to consult a professional plumber. They can diagnose the problem, recommend the appropriate course of action, and safely perform any necessary maintenance or repairs.

By understanding the different types of water heaters and their maintenance and repair requirements, you can make informed decisions about your home's hot water supply and ensure that your water heater operates efficiently and reliably for years to come.

Flushing and Cleaning Your Water Heater for Optimal Performance

Regular maintenance, such as flushing and cleaning your water heater, is essential for ensuring optimal performance, energy efficiency, and longevity of the appliance. Over time, sediment and mineral buildup can accumulate in the tank, reducing its heating capacity, causing damage, and leading to higher energy bills. In this section, we'll provide a step-by-step guide on how to flush and clean your water heater for optimal performance.

Before You Begin:
- Determine the type of water heater you have (gas, electric, or tankless) and familiarize yourself with its components and operation.
- Gather the necessary tools and materials, such as a garden hose, a bucket, gloves, and a screwdriver.
- Set aside enough time to complete the process, as it may take several hours depending on the size of your water heater and the amount of sediment buildup.

Step 1: Turn Off the Water Heater

a. For electric water heaters:
- Locate the circuit breaker for your water heater in your home's electrical panel and switch it off.

b. For gas water heaters:
- Locate the thermostat and set it to the "Pilot" or "Off" position.

Step 2: Turn Off the Cold Water Supply

a. Locate the cold water supply valve, which is usually located above the water heater.

b. Turn the valve clockwise to shut off the cold water supply to the water heater.

Step 3: Connect a Garden Hose

a. Locate the drain valve at the bottom of the water heater.

b. Connect one end of a garden hose to the drain valve and place the other end in a bucket or near a floor drain.

Step 4: Drain the Water Heater

a. Open a hot water faucet in a nearby sink or tub to allow air to enter the tank as it drains.

b. Open the drain valve on the water heater and allow the water to flow out through the garden hose.

c. Be cautious, as the water may be hot and could cause burns.

Step 5: Flush the Tank

a. Once the tank is empty, turn the cold water supply valve back on to allow fresh water to enter the tank.

b. Let the water flow through the tank and out of the garden hose for several minutes to flush out any remaining sediment.

c. Keep an eye on the water coming out of the hose; once it runs clear, the tank is clean.

Step 6: Refill the Tank

a. Close the drain valve and remove the garden hose.

b. Keep the hot water faucet open and allow the tank to refill completely.

c. Once water starts flowing from the hot water faucet, close it.

Step 7: Restore Power and Relight the Pilot (if necessary)

a. For electric water heaters:

- Switch the circuit breaker back on to restore power to the water heater.

b. For gas water heaters:

- If your water heater has a pilot light, relight it following the manufacturer's instructions.
- Turn the thermostat back to its original temperature setting.

Step 8: Check for Leaks and Proper Operation

a. Inspect the water heater and the surrounding area for any leaks or signs of damage.

b. Test the hot water at a nearby faucet to ensure that the water heater is functioning properly.

Tips for Maintaining Your Water Heater:

- Flush and clean your water heater annually to prevent sediment buildup and maintain efficiency.
- Test the pressure relief valve regularly to ensure it's functioning properly and to prevent dangerous pressure buildup.
- Inspect the anode rod every few years and replace it if it's badly corroded to prevent tank damage.
- Insulate your water heater and pipes to reduce heat loss and improve energy efficiency.
- Consider installing a water softener if you have hard water to minimize mineral buildup in your water heater and plumbing.

By following these steps and maintaining your water heater regularly, you can ensure optimal performance, energy efficiency, and a longer lifespan for your appliance. If you're unsure about any aspect of the flushing and cleaning process or encounter any issues, don't hesitate to consult a professional plumber for guidance and assistance.

Replacing a Water Heater Element or Thermostat

If your electric water heater is not heating water properly, the problem may lie with a faulty heating element or thermostat. These components are essential for the proper functioning of your water heater, and replacing them is a common repair task. In this section, we'll provide a step-by-step guide on how to replace a water heater element or thermostat.

Before You Begin:

- Make sure your water heater is an electric model, as this guide does not apply to gas water heaters.
- Gather the necessary tools and materials, such as a screwdriver, an element wrench, a multimeter, and a replacement element or thermostat.
- Familiarize yourself with the water heater's components and the location of the elements and thermostats.

Step 1: Turn Off the Power Supply
a. Locate the circuit breaker for your water heater in your home's electrical panel.
b. Switch off the circuit breaker to cut power to the water heater completely.

Step 2: Drain the Water Heater
a. Turn off the cold water supply valve, usually located above the water heater.
b. Connect a garden hose to the drain valve at the bottom of the water heater and place the other end in a bucket or near a floor drain.
c. Open a hot water faucet in a nearby sink or tub to allow air to enter the tank as it drains.
d. Open the drain valve and let the water heater drain completely to avoid burns and spills when removing the element or thermostat.

71

Replacing a Heating Element:

Step 3: Remove the Access Panel and Insulation
a. Locate the access panel for the element you need to replace (upper or lower).
b. Remove the screws holding the access panel in place and set them aside.
c. Carefully remove the insulation and plastic cover, if present.

Step 4: Disconnect the Wiring
a. Note the wiring configuration and take a picture if necessary to ensure proper reconnection.
b. Using a multimeter, double-check that the power is off by testing for voltage at the element terminals.
c. Disconnect the wires from the element terminals.

Step 5: Remove the Old Element
a. Using an element wrench, unscrew the old element counterclockwise and remove it from the water heater.
b. Inspect the tank opening for any damage or debris, and clean it if necessary.

Step 6: Install the New Element
a. Wrap the threads of the new element with Teflon tape to ensure a tight, leak-free seal.
b. Insert the new element into the tank opening and tighten it clockwise using the element wrench.

Step 7: Reconnect the Wiring and Reassemble
a. Reconnect the wires to the element terminals, ensuring the proper configuration.
b. Replace the plastic cover and insulation, if present.
c. Reinstall the access panel and secure it with the screws.

Replacing a Thermostat:

Step 3: Locate the Thermostat
a. The thermostat is usually located behind the access panel, above the corresponding heating element.

Step 4: Disconnect the Wiring
a. Note the wiring configuration and take a picture if necessary to ensure proper reconnection.
b. Using a multimeter, double-check that the power is off by testing for voltage at the thermostat terminals.
c. Disconnect the wires from the thermostat terminals.

Step 5: Remove the Old Thermostat
a. Remove the screws or clips holding the thermostat in place.
b. Carefully remove the old thermostat from the water heater.

Step 6: Install the New Thermostat
a. Position the new thermostat in the same location as the old one.
b. Secure the new thermostat in place using the screws or clips.

Step 7: Reconnect the Wiring and Reassemble
a. Reconnect the wires to the thermostat terminals, ensuring the proper configuration.
b. Replace the plastic cover and insulation, if present.
c. Reinstall the access panel and secure it with the screws.

Step 8: Refill the Water Heater and Restore Power
a. Close the drain valve and remove the garden hose.
b. Turn on the cold water supply valve to refill the water heater.
c. Once the tank is full, turn the circuit breaker back on to restore power to the water heater.

Step 9: Test and Adjust

a. Allow the water heater to heat up for about an hour.

b. Test the hot water at a nearby faucet to ensure the water heater is functioning properly.

c. If necessary, adjust the temperature setting on the thermostat to your desired level.

By following these steps, you can successfully replace a faulty heating element or thermostat in your electric water heater. Regular maintenance, such as flushing the tank and checking the anode rod, can help extend the life of your water heater and its components.

If you're unsure about any aspect of the replacement process or encounter issues, don't hesitate to consult a professional plumber for guidance and assistance. Always prioritize safety when working with electrical components and hot water systems.

Chapter 7
Preventing and Dealing with Frozen Pipes

Insulating Your Pipes to Prevent Freezing

Frozen pipes can be a major problem for homeowners, especially during cold winter months. When water freezes inside pipes, it expands and can cause the pipes to crack or burst, leading to costly repairs and water damage. In this section, we'll discuss how to prevent frozen pipes by insulating them properly and what to do if your pipes do freeze.

Insulating Your Pipes to Prevent Freezing:

Step 1: Identify Pipes at Risk of Freezing
a. Locate pipes in unheated areas of your home, such as attics, crawl spaces, and exterior walls.
b. Identify pipes that are exposed to cold drafts, such as those near windows or doors.
c. Pay special attention to pipes that have frozen in the past, as they are more likely to freeze again.

Step 2: Choose the Right Insulation Material
a. Foam pipe sleeves: These are the most common type of pipe insulation. They are easy to install and come in various sizes to fit different pipe diameters.
b. Fiberglass pipe wrap: This type of insulation is more versatile and can be wrapped around pipes of various sizes and shapes. It is also fire-resistant.
c. Heat tape: This is an electrical heating element that can be wrapped around pipes to keep them warm. It is particularly useful for pipes that are prone to freezing or difficult to insulate.

Step 3: Prepare the Pipes for Insulation

a. Clean the pipes thoroughly to remove any dirt, grease, or debris that could prevent the insulation from adhering properly.

b. Repair any leaks or cracks in the pipes before insulating them, as insulation will not prevent water from escaping through existing leaks.

Step 4: Install the Insulation

a. For foam pipe sleeves:

- Measure the length of the pipe you need to insulate and cut the sleeve to the appropriate length using a utility knife.
- Slip the sleeve over the pipe, making sure it fits snugly. If the sleeve has a slit, make sure the slit is facing downward to prevent water from seeping in.
- Secure the sleeve in place with duct tape or zip ties, if necessary.

b. For fiberglass pipe wrap:

- Measure the length of the pipe you need to insulate and cut the wrap to the appropriate length using scissors.
- Wrap the insulation around the pipe, making sure to overlap the edges by at least 1 inch.
- Secure the wrap in place with duct tape or zip ties, making sure the tape or ties are snug but not too tight, as this could compress the insulation and reduce its effectiveness.

c. For heat tape:

- Follow the manufacturer's instructions carefully, as improper installation can be a fire hazard.
- Make sure the heat tape is approved for use on pipes and is rated for the appropriate pipe diameter and length.
- Do not overlap the heat tape, as this can cause overheating and damage to the pipes.

Step 5: Seal and Protect the Insulation

a. Use a weatherproof sealant, such as silicone caulk, to seal any gaps or seams in the insulation to prevent cold air from reaching the pipes.

b. In areas where the insulation may be exposed to physical damage, such as in a crawl space, consider installing a protective cover, such as a PVC pipe or a specially designed insulation jacket.

Dealing with Frozen Pipes:

If you turn on a faucet and only a trickle of water comes out, or no water at all, you may have a frozen pipe. Here's what to do:

Step 1: Locate the Frozen Pipe

a. Check all faucets in your home to determine which ones are affected by the frozen pipe.

b. Follow the pipe back from the affected faucet to locate the area where it is most likely to be frozen. This is often in an unheated area or near an exterior wall.

Step 2: Thaw the Frozen Pipe

a. Keep the affected faucet open, as running water will help melt the ice and indicate when the pipe is thawed.

b. Apply heat to the frozen section of the pipe using one of the following methods:

- Wrap the pipe in a heating pad or electric heat tape.
- Use a hair dryer or portable space heater to blow warm air onto the pipe.
- Place a towel soaked in hot water around the pipe.

c. Never use an open flame, such as a blowtorch or a candle, to thaw a frozen pipe, as this can be a fire hazard and may damage the pipe.

d. Start thawing the pipe near the faucet and work your way back toward the frozen section. This allows the melting ice to escape through the open faucet.

Step 3: Check for Leaks

a. Once the pipe is thawed and water is flowing freely, carefully inspect the pipe for any cracks or leaks that may have been caused by the freezing.

b. If you find a leak, turn off the main water supply to your home immediately and call a plumber for repairs.

Step 4: Prevent Future Freezing

a. After thawing the pipe, take steps to prevent it from freezing again by insulating it properly, as described in the previous section.

b. Consider leaving the affected faucet slightly open during extremely cold weather to keep water moving through the pipe and prevent freezing.

By following these steps to insulate your pipes and knowing how to deal with frozen pipes, you can prevent costly damage and ensure a reliable water supply throughout the winter months. Remember, if you are unsure about any aspect of pipe insulation or dealing with frozen pipes, it is always best to consult a professional plumber for guidance and assistance.

Thawing Frozen Pipes Safely and Effectively

When pipes freeze, it's essential to act quickly and thaw them safely to prevent damage and restore water flow. In this section, we'll provide a detailed, step-by-step guide on how to thaw frozen pipes safely and effectively.

Before You Begin:
- Locate the frozen pipe and determine the extent of the freezing. Check all faucets in your home to identify which ones are affected.
- Turn off the main water supply to your home to prevent water from escaping if the pipe has cracked or burst.
- Gather necessary materials, such as a hair dryer, space heater, or heating pad.

Step 1: Open the Affected Faucets
a. Turn on all the faucets connected to the frozen pipe, both hot and cold.
b. This will help relieve pressure in the pipe and allow water to escape as the ice melts.

Step 2: Apply Heat to the Frozen Section
a. Start thawing the pipe near the faucet and work your way toward the frozen section. This allows the melting ice to escape through the open faucet.
b. Use one of the following methods to apply heat:
- Hair dryer: Blow warm air directly onto the frozen section of the pipe, keeping the dryer moving to avoid overheating any one spot.
- Space heater: Position a portable space heater near the frozen pipe, making sure to keep it at a safe distance to prevent fire hazards.

- Heating pad: Wrap a heating pad around the frozen section of the pipe, securing it in place with duct tape or string.
- Hot towels: Soak towels in hot water and wrap them around the frozen pipe. Replace the towels as they cool.

c. Continue applying heat until water begins to flow freely from the faucets.

Step 3: Monitor the Thawing Process

a. Keep an eye on the pipe as it thaws to ensure that the heating method you're using isn't causing any damage to the pipe or creating a fire hazard.

b. If you're using an electrical device like a hair dryer or space heater, be cautious of any water that may be present to avoid electrical shock.

c. Never leave a heating device unattended, and make sure to keep it away from flammable materials.

Step 4: Check for Leaks

a. Once the pipe has thawed and water is flowing freely, carefully inspect the pipe for any cracks or leaks that may have been caused by the freezing.

b. If you find a leak, turn off the main water supply immediately and call a plumber for repairs.

Step 5: Insulate the Pipe to Prevent Future Freezing

a. After thawing the pipe, take steps to prevent it from freezing again by insulating it properly.

b. Use foam pipe sleeves, fiberglass pipe wrap, or heat tape to insulate the pipe, paying special attention to areas that are prone to freezing.

Safety Precautions:

Never use an open flame, such as a blowtorch, kerosene or propane heater, or a charcoal stove, to thaw a frozen pipe. These methods can be extremely dangerous and may cause fires or damage the pipe.

- Avoid using electrical devices near standing water to prevent the risk of electrical shock.
- If you suspect a pipe has burst or if you're unable to locate the frozen section, call a plumber immediately.
- If the frozen pipe is located in an area that you can't easily access, such as behind a wall or in a crawl space, it's best to call a professional plumber to handle the situation.

Tips to Prevent Frozen Pipes:

- Keep your home's thermostat set to a minimum of 55°F (13°C), even when you're away.
- Open kitchen and bathroom cabinet doors to allow warm air to circulate around the pipes.
- Let cold water drip from faucets served by exposed pipes during extremely cold weather.
- Seal leaks and drafts in your home to keep cold air away from pipes.
- Consider installing a freeze alarm that will alert you if the temperature in your home drops below a preset level.

By following these steps and taking the necessary safety precautions, you can thaw frozen pipes safely and effectively, minimizing the risk of damage and restoring water flow to your home. Remember, if you're ever unsure about how to handle a frozen pipe situation or if you suspect more extensive damage, it's always best to call a professional plumber for assistance.

Repairing Burst Pipes and Water Damage

Burst pipes can cause significant water damage to your home, leading to costly repairs and potential health hazards. In this section, we'll provide a detailed guide on how to repair burst pipes and deal with water damage.

Step 1: Shut Off the Water Supply
a. Locate your home's main water shut-off valve and turn it off immediately to stop water from flowing through the damaged pipe.
b. If you can't locate the main shut-off valve, or if it's not working, call your local water utility company for assistance.

Step 2: Drain the Pipes
a. Open all faucets in your home to drain the remaining water from the pipes.
b. Flush all toilets to remove water from their tanks and bowls.
c. If your home has a water heater, turn off its power supply (electricity or gas) and drain it to prevent damage.

Step 3: Assess the Damage
a. Locate the burst pipe and assess the extent of the damage.
b. Check for any visible cracks, splits, or holes in the pipe.
c. Determine the type of pipe material (e.g., copper, PVC, or PEX) and the size of the damaged section.

Step 4: Repair the Burst Pipe (for minor damage)
a. If the damage is minor, such as a small crack or pinhole leak, you may be able to repair it temporarily using a pipe repair kit or pipe repair tape.
b. Clean the damaged area of the pipe thoroughly and dry it completely.
c. Apply the repair kit or tape according to the manufacturer's instructions, making sure to cover the damaged area completely.

d. Keep in mind that these repairs are temporary and that you'll need to have the pipe replaced as soon as possible.

Step 5: Replace the Burst Pipe (for major damage)
a. If the damage is severe, or if the temporary repair fails, you'll need to replace the damaged section of the pipe.
b. Measure the length of the damaged section and add a few inches on each end for the replacement pipe.
c. Cut out the damaged section using a pipe cutter or a hacksaw, making sure to cut straight and clean edges.
d. Clean the ends of the remaining pipe and the replacement section, and dry them completely.
e. Install the replacement section using the appropriate fittings (e.g., couplings, unions, or elbows) and sealants (e.g., solder, adhesive, or thread seal tape).
f. Turn the water supply back on and check for leaks. If any leaks are found, turn the water off immediately and repair the leaks before proceeding.

Step 6: Clean Up Water Damage
a. Remove standing water using a wet/dry vacuum, mops, or towels.
b. Use fans and dehumidifiers to dry out the affected areas thoroughly. This can take several days, depending on the extent of the damage.
c. Discard any water-damaged items that cannot be salvaged, such as carpet padding, drywall, or insulation.
d. Clean and disinfect all surfaces that came into contact with the water to prevent mold and mildew growth.

Step 7: Prevent Future Burst Pipes
a. Insulate pipes in unheated areas, such as attics, crawl spaces, and exterior walls, to prevent them from freezing.

b. Seal any cracks or gaps in your home's foundation or exterior walls to keep cold air away from pipes.

c. Disconnect and drain outdoor hoses during the winter months.

d. Keep your home's thermostat set to a minimum of 55°F (13°C), even when you're away.

When to Call a Professional:

- If you're unsure about any aspect of the repair process or don't have the necessary tools and skills.
- If the damage is extensive or if you suspect that there may be hidden water damage.
- If you're unable to locate the source of the leak or if the leak is in a difficult-to-reach location.
- If you suspect that there may be a risk of contamination from sewage or other hazardous materials.

In these cases, it's best to call a professional plumber and a water damage restoration company to handle the situation safely and effectively.

By following these steps and taking the necessary precautions, you can repair burst pipes and minimize water damage to your home. Remember, acting quickly and thoroughly is essential to prevent further damage and potential health hazards.

Chapter 8
Advanced Plumbing Techniques and Projects

Installing a New Dishwasher or Garbage Disposal

Burst pipes can cause significant water damage to your home, leading to costly repairs and potential health hazards. In this section, we'll provide a detailed guide on how to repair burst pipes and deal with water damage.

Step 1: Shut Off the Water Supply

a. Locate your home's main water shut-off valve and turn it off immediately to stop water from flowing through the damaged pipe.
b. If you can't locate the main shut-off valve, or if it's not working, call your local water utility company for assistance.

Step 2: Drain the Pipes

a. Open all faucets in your home to drain the remaining water from the pipes.
b. Flush all toilets to remove water from their tanks and bowls.
c. If your home has a water heater, turn off its power supply (electricity or gas) and drain it to prevent damage.

Step 3: Assess the Damage

a. Locate the burst pipe and assess the extent of the damage.
b. Check for any visible cracks, splits, or holes in the pipe.
c. Determine the type of pipe material (e.g., copper, PVC, or PEX) and the size of the damaged section.

Step 4: Repair the Burst Pipe (for minor damage)

a. If the damage is minor, such as a small crack or pinhole leak, you may be able to repair it temporarily using a pipe repair kit or pipe repair tape.

b. Clean the damaged area of the pipe thoroughly and dry it completely.

c. Apply the repair kit or tape according to the manufacturer's instructions, making sure to cover the damaged area completely.

d. Keep in mind that these repairs are temporary and that you'll need to have the pipe replaced as soon as possible.

Step 5: Replace the Burst Pipe (for major damage)

a. If the damage is severe, or if the temporary repair fails, you'll need to replace the damaged section of the pipe.

b. Measure the length of the damaged section and add a few inches on each end for the replacement pipe.

c. Cut out the damaged section using a pipe cutter or a hacksaw, making sure to cut straight and clean edges.

d. Clean the ends of the remaining pipe and the replacement section, and dry them completely.

e. Install the replacement section using the appropriate fittings (e.g., couplings, unions, or elbows) and sealants (e.g., solder, adhesive, or thread seal tape).

f. Turn the water supply back on and check for leaks. If any leaks are found, turn the water off immediately and repair the leaks before proceeding.

Step 6: Clean Up Water Damage

a. Remove standing water using a wet/dry vacuum, mops, or towels.

b. Use fans and dehumidifiers to dry out the affected areas thoroughly. This can take several days, depending on the extent of the damage.

c. Discard any water-damaged items that cannot be salvaged, such as carpet padding, drywall, or insulation.

d. Clean and disinfect all surfaces that came into contact with the water to prevent mold and mildew growth.

Step 7: Prevent Future Burst Pipes

a. Insulate pipes in unheated areas, such as attics, crawl spaces, and exterior walls, to prevent them from freezing.

b. Seal any cracks or gaps in your home's foundation or exterior walls to keep cold air away from pipes.

c. Disconnect and drain outdoor hoses during the winter months.

d. Keep your home's thermostat set to a minimum of 55°F (13°C), even when you're away.

When to Call a Professional:

- If you're unsure about any aspect of the repair process or don't have the necessary tools and skills.
- If the damage is extensive or if you suspect that there may be hidden water damage.
- If you're unable to locate the source of the leak or if the leak is in a difficult-to-reach location.
- If you suspect that there may be a risk of contamination from sewage or other hazardous materials.

In these cases, it's best to call a professional plumber and a water damage restoration company to handle the situation safely and effectively.

By following these steps and taking the necessary precautions, you can repair burst pipes and minimize water damage to your home. Remember, acting quickly and thoroughly is essential to prevent further damage and potential health hazards.

Replacing a Sewer Line or Main Water Valve

Replacing a sewer line or main water valve is a major plumbing project that often requires professional expertise. However, understanding the process can help you make informed decisions and work effectively with your plumber. In this section, we'll provide a detailed overview of the steps involved in replacing a sewer line or main water valve.

Replacing a Sewer Line:

Step 1: Identify the Problem
a. Signs of a damaged sewer line include frequent clogs, slow drains, gurgling sounds, sewage odors, and wet spots in your yard.
b. If you suspect a problem with your sewer line, contact a professional plumber to inspect the line using a sewer camera.

Step 2: Locate the Sewer Line
a. Sewer lines typically run from your home to the city's main sewer line or your septic tank.
b. Your plumber will use a sewer camera or other detection equipment to locate the damaged section of the line.

Step 3: Obtain Necessary Permits
a. Before starting any excavation work, your plumber will need to obtain the necessary permits from your local government.
b. The permitting process ensures that the work complies with local building codes and safety regulations.

Step 4: Excavate the Damaged Section
a. Using heavy equipment, your plumber will excavate the area around the damaged section of the sewer line.

b. The excavation may involve removing soil, concrete, or asphalt, depending on the location of the line.

Step 5: Remove the Damaged Section
a. Once the damaged section is exposed, your plumber will cut it out using a saw or other cutting tools.
b. The remaining ends of the sewer line will be cleaned and prepared for the new section.

Step 6: Install the New Sewer Line
a. Your plumber will measure and cut the new section of sewer line to fit the excavated area.
b. The new section will be connected to the existing sewer line using couplings or other fittings.
c. The joints will be sealed to prevent leaks and ensure a proper fit.

Step 7: Test the New Sewer Line
a. Before backfilling the excavated area, your plumber will test the new sewer line for leaks and proper flow.
b. This may involve running water through the line and using a sewer camera to inspect the interior.

Step 8: Backfill and Restore the Area
a. Once the new sewer line has passed inspection, your plumber will backfill the excavated area with soil.
b. Any concrete or asphalt that was removed will be replaced and smoothed to match the surrounding surface.

Replacing a Main Water Valve:

Step 1: Locate the Main Water Valve
a. The main water valve is typically located near the water meter, where the main water line enters your home.

b. If you're unsure of the location, consult your property records or contact your local water utility company.

Step 2: Shut Off the Water Supply

a. Before replacing the main water valve, you'll need to shut off the water supply to your entire home.

b. This may involve contacting your water utility company to turn off the water at the street level.

Step 3: Drain the Pipes

a. Open all faucets in your home to drain the remaining water from the pipes.

b. This will help prevent water from spilling when you remove the old valve.

Step 4: Remove the Old Valve

a. Using a wrench or other tools, your plumber will loosen the fittings on either side of the old valve.

b. The old valve will be removed, and the ends of the water line will be cleaned and prepared for the new valve.

Step 5: Install the New Valve

a. Your plumber will measure and select a new valve that matches the size and type of your water line.

b. The new valve will be positioned in place and connected to the water line using fittings and sealants.

c. The fittings will be tightened to ensure a proper, leak-free connection.

Step 6: Test the New Valve

a. Before turning the water supply back on, your plumber will test the new valve for leaks.

b. This may involve partially opening the valve and checking the fittings for any signs of water.

Step 7: Restore Water Supply and Check for Leaks

a. Once the new valve has passed inspection, your plumber will turn the water supply back on.

b. All faucets in your home will be checked for proper flow and any signs of leaks.

c. If any leaks are detected, your plumber will make the necessary adjustments to the fittings or valve.

When to Call a Professional:

- Replacing a sewer line or main water valve is a complex and labor-intensive process that requires specialized knowledge and equipment.
- Attempting to replace these components yourself can lead to costly mistakes, property damage, and potential health hazards.
- Always contact a licensed and experienced plumber to handle these types of projects to ensure that the work is done safely, efficiently, and in compliance with local building codes.

By understanding the process of replacing a sewer line or main water valve, you can work effectively with your plumber and make informed decisions about your home's plumbing system. Regular maintenance and prompt attention to any signs of damage can help prevent the need for these major repairs and keep your plumbing system functioning properly for years to come.

Adding a New Plumbing Fixture or Expanding Your System

Adding a new plumbing fixture or expanding your plumbing system can enhance the functionality and value of your home. Whether you're installing a new sink, toilet, or shower, or adding a new bathroom or laundry room, understanding the process can help you plan and execute your project successfully. In this section, we'll provide a detailed overview of the steps involved in adding a new plumbing fixture or expanding your plumbing system.

Step 1: Plan Your Project
a. Determine the scope of your project, including the type and location of the new fixture or expansion.
b. Consider factors such as the available space, existing plumbing layout, and potential obstacles.
c. Consult with a professional plumber or architect to help you develop a detailed plan and ensure that your project complies with local building codes.

Step 2: Obtain Necessary Permits
a. Before starting any plumbing work, you'll need to obtain the necessary permits from your local government.
b. The permitting process ensures that your project meets safety and building code requirements.

Step 3: Prepare the Area
a. Clear the area where the new fixture or expansion will be located.
b. Remove any existing fixtures, flooring, or drywall as needed to access the plumbing.
c. If you're expanding your system, you may need to create new openings in walls, floors, or ceilings to accommodate the new plumbing.

92

Step 4: Install the Rough-In Plumbing

a. The rough-in plumbing includes the water supply lines, drain pipes, and vent stacks that will connect to your new fixture or expansion.

b. If you're adding a new fixture, your plumber will tap into the existing water supply and drainage lines and run new pipes to the location of the fixture.

c. If you're expanding your system, your plumber will install new main supply and drainage lines and connect them to the existing system.

d. The rough-in plumbing will be tested for leaks and proper flow before proceeding to the next step.

Step 5: Install the Fixture

a. Once the rough-in plumbing is complete, your plumber will install the new fixture according to the manufacturer's instructions.

b. This may involve connecting the fixture to the water supply and drainage lines, sealing the connections, and securing the fixture in place.

c. If you're installing a new sink or toilet, your plumber will also install the necessary faucets, valves, and other components.

Step 6: Test the New Fixture or System

a. After the installation is complete, your plumber will test the new fixture or system for proper operation.

b. This may involve running water through the fixture, checking for leaks, and ensuring that the drainage is flowing properly.

c. Any necessary adjustments will be made to ensure that the fixture or system is functioning optimally.

Step 7: Finish the Area

a. Once the new fixture or expansion has passed inspection, you can finish the surrounding area.

b. This may involve installing new flooring, drywall, or tile, and painting or decorating the space to match the rest of your home.

c. If you've expanded your system, you may also need to update your home's electrical and ventilation systems to accommodate the new plumbing.

Tips for a Successful Project:

- Work with a licensed and experienced plumber to ensure that your project is completed safely and efficiently.
- Choose high-quality fixtures and materials that are appropriate for your home's style and needs.
- Consider the long-term maintenance and repair needs of your new fixture or expansion, and plan accordingly.
- If you're expanding your system, think about how the new plumbing will integrate with your home's existing layout and design.
- Be prepared for unexpected challenges or delays, and budget accordingly.

By following these steps and tips, you can successfully add a new plumbing fixture or expand your plumbing system, enhancing the functionality and value of your home. Remember to always prioritize safety and quality, and don't hesitate to seek professional guidance when needed.

Conclusion

Throughout this book, we've explored the intricate world of home plumbing, from understanding the basic components of your plumbing system to tackling a wide range of common plumbing issues. By now, you should have a solid foundation of knowledge and practical skills that will empower you to take on most plumbing challenges with confidence and competence.

We've covered everything from fixing leaky faucets and unclogging drains to replacing toilets and water heaters. You've learned how to diagnose problems, select the right tools and materials, and follow step-by-step procedures to resolve issues safely and effectively. Along the way, you've also gained valuable insights into how to maintain your plumbing system, prevent future problems, and know when it's time to call in a professional.

But mastering home plumbing is about more than just acquiring technical skills. It's about developing a mindset of self-reliance, resourcefulness, and perseverance. It's about being willing to get your hands dirty, learn from your mistakes, and take pride in your accomplishments. And it's about gaining a deeper appreciation for the complex and essential role that plumbing plays in our daily lives.

As you continue on your journey as a DIY plumber, remember that the key to success is always to prioritize safety, plan carefully, and never stop learning. Keep expanding your knowledge by staying up-to-date with the latest plumbing techniques, technologies, and best practices. Engage with the wider community of DIY enthusiasts and professionals, and don't be afraid to ask for help or advice when you need it.

With the knowledge and skills you've gained from this book, you're well-equipped to handle most plumbing issues that come your way. But remember, there will always be new challenges and opportunities to grow. Embrace them with the same curiosity, determination, and sense of adventure that brought you this far.

In the end, mastering home plumbing is about more than just fixing pipes and saving money. It's about taking control of your home, your life, and your future. It's about building the confidence and self-reliance that will serve you well in all aspects of your life. And it's about discovering the deep satisfaction that comes from solving problems, overcoming obstacles, and making a tangible difference in your world.

So go forth with courage, armed with the knowledge and skills you've gained from this book. Tackle those plumbing challenges head-on, and never stop learning, growing, and improving. With persistence, patience, and a little bit of plumbing prowess, there's no limit to what you can achieve.

Happy plumbing!